Electronics For Absolute Beginners

A Beginner's Guide To Understanding Electronics

Copyright@2023

Jace Pollux

Table of content

Introduction 4

    A. Electronics 4

    B. Basic Tools And Components For Electronics 9

Chapter 1: The Fundamentals Of Electricity 13

    A. Understanding Electricity And Current Flow 13

    B. Voltage, Current, And Resistance ... 18

    C. Ohm's Law And Its Applications 23

    D. Introduction To Circuit Diagrams ... 28

Chapter 2: Passive Components 33

    A. Resistors: Types, Color Coding, And Applications 33

    B. Capacitors: Types, Capacitance, And Charging/Discharging 40

    C. Inductors: Basics And Inductance ... 48

    D. Understanding Passive Component Values 54

Chapter 3: Active Components 58

    A. Introduction To Semiconductors 58

B. Diodes: Types And Applications.....64

C. Transistors: Bipolar Junction Transistors (BJTs) and Field-Effect Transistors (FETs)...............71

D. Integrated Circuits (ICs) And Their Uses 77

Chapter 4: Circuit Analysis.......................83

A. Series And Parallel Circuits 83

B. Kirchhoff's Laws And Their Application 88

C. Voltage And Current Division.........92

D. Introduction To Breadboarding.......96

Chapter 5: Basic Electronic Devices And Applications 101

A. Light-Emitting Diodes (LEDs)......101

B. Switches And Push Buttons...........107

C. Buzzer And Speaker Circuits 112

D. Introduction To Sensors: Light, Temperature, And Motion...................116

Chapter 6: Power Supplies And Voltage Regulation 122

A. Types Of Power Supplies: Batteries, Ac Adapters, And Voltage Regulators122
B. Voltage Regulation Using Zener Diodes 127
C. Introduction To Linear And Switching Regulators 131

Chapter 7: Integrated Circuits And Microcontrollers 137
A. Introduction To IC Families: TTL, CMOS, And Op-Amps 137
Microcontrollers: Arduino And Raspberry Pi 142

Chapter 8: Soldering And Prototyping 147
A. Introduction To Soldering Tools And Techniques 147
B. Building And Assembling Electronic Circuits 153
C. Prototyping On Breadboards And PCBs 158

Chapter 9: Troubleshooting And Maintenance 165

A. Common Electronics Problems And Solutions 165
B. Using Multimeters And Oscilloscopes For Testing 171
C. Safety Precautions In Electronics .. 177

Introduction

A. Electronics

Electronics is the field of study and application that deals with the control, manipulation, and transmission of electrical signals to perform various functions. It encompasses the design, development, and use of electronic devices, circuits, and systems.

Electronics plays a vital role in modern society, powering a wide range of devices and technologies we rely on daily. From smartphones and computers to televisions, medical equipment, transportation systems, and more, electronics are integral to our lives.

Key Concepts in Electronics:

1. Circuits: Circuits are the building blocks of electronic systems. They consist of interconnected components such as resistors, capacitors, transistors, and integrated circuits that allow the flow of electrical current to perform specific functions.

2. Components: Electronic components are the basic units used in circuits. These include resistors, capacitors, inductors, diodes, transistors, integrated circuits, sensors, and more. Each component has a specific function and properties that contribute to the overall operation of the circuit.

3. Signals: Electronics deals with the manipulation and processing of electrical signals. Signals can be analog or digital and carry information or data. Signal processing techniques are employed to modify, amplify, filter, or convert signals to suit specific applications.

4. Digital Electronics: Digital electronics involve the use of discrete voltage levels (0s and 1s) to represent and process information. It forms the basis of digital computers, microcontrollers, and other digital systems. Boolean algebra and logic gates are fundamental to digital electronics.

5. Integrated Circuits (ICs): Integrated circuits are miniaturized electronic circuits fabricated on a single chip of semiconductor material. They can contain thousands to millions of transistors and other components, allowing for complex functionality in a small form factor.

6. Power Electronics: Power electronics deals with the conversion and control of electrical power. It involves the design of circuits and systems for applications such as power supplies, motor drives, renewable energy systems, and electric vehicles.

7. Communication Systems: Electronics plays a significant role in communication systems, enabling the transmission and reception of signals over long distances. It encompasses areas such as telecommunication, wireless communication, satellite systems, and data networking.

8. Sensors and Actuators: Sensors detect physical, chemical, or environmental parameters and convert them into electrical signals. Actuators, on the other hand, convert electrical signals into physical actions or control mechanisms. They are essential in applications such as automation, robotics, and Internet of Things (IoT) devices.

B. Basic Tools And Components For Electronics

When it comes to working with electronics, there are several basic tools and components that are commonly used. Here's a list of some essential tools and components:

Basic Tools:
1. Soldering Iron: Used for soldering components onto circuit boards or joining wires together.
2. Wire Cutters: Used to cut and strip wires to the desired length.
3. Needle-Nose Pliers: Useful for bending wires, holding small components, or reaching into tight spaces.
4. Screwdrivers: Assorted screwdrivers for tightening or loosening screws in electronic devices.

5. Multimeter: In electronic circuits, multimeters are employed to gauge voltage, current, and resistance levels.

6. Breadboard: A reusable platform for prototyping and testing circuits without the need for soldering.

7. Tweezers: Used for handling small electronic components or placing them precisely on a circuit board.

8. Wire Strippers: Specialized tools for cleanly removing insulation from wires.

9. Desoldering Pump or Desoldering Braid: Used to remove solder and desolder components.

10. Helping Hands: A tool with alligator clips to hold components or wires in place during soldering or assembly.

Basic Components:

1. Resistors: Used to limit or control the flow of electric current in a circuit.

2. Capacitors: Store and release electrical energy, used for filtering and timing applications.

3. Diodes: They permit the passage of electric current in one direction while inhibiting it from flowing in the opposite direction.

4. LEDs (Light-Emitting Diodes): Emit light when current flows through them, often used as indicators.

5. Transistors: Act as amplifiers or electronic switches, controlling the flow of current in a circuit.

6. Integrated Circuits (ICs): Miniature electronic circuits containing many components such as microprocessors or memory chips.

7. Potentiometers: Variable resistors used to control voltage or adjust levels like volume or brightness.

8. Inductors: Store energy in a magnetic field, commonly used in filters and oscillators.

9. Switches: Control the flow of current by opening or closing a circuit.

10. Batteries: Provide portable power for electronic devices.

It's important to note that this is not an exhaustive list, as electronics is a vast field with numerous tools and components available. The specific tools and components required will depend on the project or task at hand.

Chapter 1: The Fundamentals Of Electricity

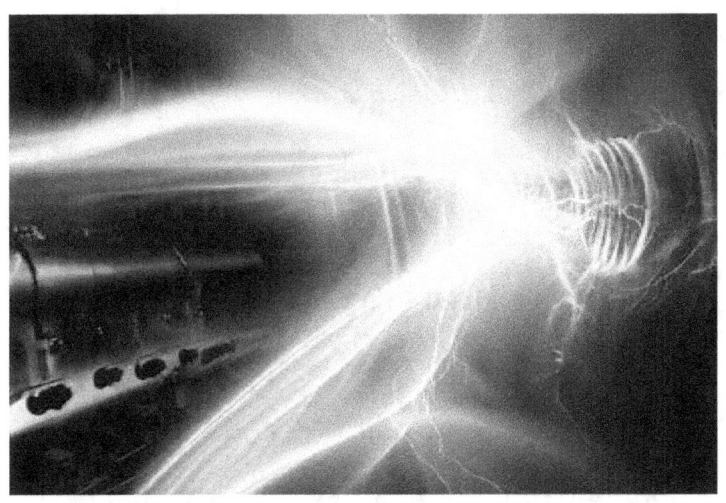

A. Understanding Electricity And Current Flow

Understanding electricity and current flow is fundamental to comprehending how electronics work. Here's a basic explanation:

Electricity: Electricity is a type of energy that arises from the motion/movement of charged particles, like electrons. It is typically generated through various means, including power plants, batteries, solar panels, and more. Electricity can be harnessed and utilized to power electrical devices and perform work.

Current: Electric current refers to the flow of electric charge in a circuit. It is measured in units called amperes (A) and is denoted by the symbol "I." Current can flow through conductive materials like wires, which provide a path for the movement of charged particles. The flow of current is driven by a voltage difference or potential difference (measured in volts, V) between two points in a circuit.

Types of Current:
1. Direct Current (DC): In a DC circuit, the flow of electric charge is unidirectional, meaning it flows in one direction only. The

voltage polarity remains constant over time. Batteries and many electronic devices, such as mobile phones and laptops, operate on DC power.

2. Alternating Current (AC): In an AC circuit, the flow of electric charge periodically reverses direction. It oscillates back and forth, typically at a specific frequency, often measured in hertz (Hz). AC is the type of current supplied by power grids and is commonly used in household electrical systems.

Components of Current Flow:
1. Voltage: Voltage, also known as electric potential difference, represents the driving force that pushes electric charge through a circuit. It is measured in volts (V) and denoted by the symbol "V." Voltage can be

visualized as the pressure or "push" that causes the flow of electrons.

2. Resistance: Resistance (measured in ohms, Ω) is a property of a material or component that opposes the flow of current. It limits the movement/flow of electrons. Materials with high resistance impede current flow, while materials with low resistance allow for easier flow. Resistors are components specifically designed to provide a specific amount of resistance in a circuit.

3. Conductors and Insulators: Conductors are materials that have low resistance and allow for easy flow of electric current. Instances encompass materials such as copper and aluminum, which are classified as metals. Insulators, on the other hand, have high resistance and inhibit the flow of current. Materials like rubber, plastic, and wood are commonly used as insulators to prevent accidental electric shock or short circuits.

Ohm's Law: Ohm's Law is a fundamental principle that relates voltage, current, and resistance in a circuit. Ohm's Law asserts that the electric current passing through a conductor is directly proportional to the voltage applied across it and inversely proportional to the resistance of the conductor. Mathematically, Ohm's Law is expressed as $I = V/R$, where I is the current, V is the voltage, and R is the resistance.

B. Voltage, Current, And Resistance

Voltage, current, and resistance are fundamental concepts in electricity and electronics. Now, let's delve deeper into each of these concepts to gain a better understanding of them:

Voltage:

Voltage, also known as electric potential difference, is a measure of the electrical potential energy per unit charge between two points in an electrical circuit. It represents the "push" or force that drives electric charge to flow. Voltage is measured in volts (V) and is denoted by the symbol "V."

Voltage can be visualized as a difference in electric potential between two points in a circuit. It creates an electric field that exerts a force on charged particles, causing them to move from a higher voltage point (positive terminal) to a lower voltage point (negative terminal).

In practical terms, voltage is what provides the energy to power electrical devices and make them function. Batteries, power outlets, and generators are sources of voltage in electrical circuits.

Current:
Current refers to the movement or flow of electric charge within an electrical circuit. It represents the rate at which charges (usually electrons) move through a conductor.
Current is measured in amperes (A) and is denoted by the symbol "I."

Current flows when there is a closed loop or path for the charges to move. It is caused by the movement of charged particles in response to the applied voltage. Current can flow in two forms:

1. Direct Current (DC): In a DC circuit, the flow of electric charge is unidirectional, meaning it flows in one direction only. The current remains constant in magnitude and direction over time. Batteries and many electronic devices operate on DC power.

2. Alternating Current (AC): In an AC circuit, the flow of electric charge periodically changes direction. It oscillates back and forth, typically at a specific frequency. AC is the type of current supplied by power grids and is commonly used in household electrical systems.

Resistance:

Resistance is a property of a material or component that opposes the flow of electric current. It restricts the movement of charged particles, converting electrical energy into other forms such as heat. Resistance is measured in ohms (Ω) and is denoted by the symbol "R."

Resistance is influenced by various factors, including the dimensions, material, and temperature of a conductor. Materials with high resistance impede the flow of current, while materials with low resistance allow for easier flow. A component called a resistor is specifically designed to provide a known amount of resistance in a circuit.

Ohm's Law:

Ohm's Law establishes a fundamental correlation between voltage, current, and resistance. According to Ohm's Law, the relationship between the current flowing through a conductor, the voltage applied across it, and the resistance of the conductor is such that the current is directly proportional to the voltage and inversely proportional to the resistance.

Mathematically, Ohm's Law is expressed as:

$$I = V / R$$

Where:
- I represents the current in amperes (A).
- V represents the voltage in volts (V).
- R represents the resistance in ohms (Ω).

Ohm's Law allows us to calculate one of the three variables (voltage, current, or resistance) when the other two are known. It is a fundamental tool for analyzing and designing electrical and electronic circuits.

C. Ohm's Law And Its Applications

Ohm's Law is a fundamental principle in electricity and electronics that relates voltage, current, and resistance in a circuit. It provides a mathematical relationship that helps analyze and design electrical systems. Let's explore Ohm's Law and its applications:

According to Ohm's Law, the electric current passing through a conductor is directly proportional to the voltage applied across it and inversely proportional to the resistance of the conductor. Mathematically, Ohm's Law is expressed as:

$I = V / R$

Where:
- I represents the current in amperes (A).
- V represents the voltage in volts (V).

- R represents the resistance in ohms (Ω).

Applications of Ohm's Law:

1. Calculating Current: Ohm's Law allows you to calculate the current flowing through a circuit when the voltage and resistance are known. Simply rearrange the formula to solve for current:

$I = V / R$

2. Determining Voltage: Ohm's Law can be used to calculate the voltage across a component or circuit when the current and resistance are known. Rearrange the formula to solve for voltage:

$V = I * R$

3. Finding Resistance: When the values of voltage and current are known in a circuit, Ohm's Law can be utilized to ascertain the resistance present in the circuit. Rearrange the formula to solve for resistance:

$R = V / I$

4. Series Circuits: Ohm's Law is applicable to series circuits, where components are connected in a single path. In a series circuit, the total resistance is determined by adding together the individual resistances that are connected in series. Using Ohm's Law, you can calculate the current flowing through each component and the voltage across each component.

5. Parallel Circuits: Ohm's Law can be used in parallel circuits, where components are connected across multiple paths. In a parallel circuit, the voltage across each component remains constant, while the total current flowing through the circuit is the sum of the individual currents flowing through each component. Ohm's Law helps calculate the current flowing through each branch and the total resistance.

6. Circuit Analysis: Ohm's Law is a powerful tool for circuit analysis. By applying Ohm's Law to various parts of a circuit, you can determine voltage drops across components, current flows through branches, and overall circuit behavior.

7. Designing Circuits: Ohm's Law is crucial in designing circuits. By understanding the relationships between voltage, current, and resistance, you can select appropriate components, determine power requirements, and ensure safe operation of the circuit.

8. Troubleshooting: Ohm's Law assists in troubleshooting electrical and electronic systems. By measuring voltage and current at different points in a circuit and comparing them with expected values calculated using Ohm's Law, you can identify faulty components or incorrect connections.

Ohm's Law is a fundamental principle that enables engineers, technicians, and enthusiasts to analyze, design, and troubleshoot electrical and electronic circuits. Its applications extend beyond simple calculations, forming the basis for advanced topics in electronics, such as circuit analysis, power calculations, and impedance matching.

D. Introduction To Circuit Diagrams

A circuit diagram, also known as a schematic diagram or electrical diagram, is a graphical representation of an electrical circuit. It uses standardized symbols to depict the components and connections within a circuit. Circuit diagrams are essential for understanding, designing, and documenting electrical and electronic systems. Here's an introduction to circuit diagrams:

Components of a Circuit Diagram:
1. Symbols: Circuit diagrams use symbols to represent different electrical and electronic components. These symbols are standardized and universally recognized, allowing for easy interpretation. Examples of common symbols include resistors,

capacitors, batteries, switches, transistors, and more.

2. Lines and Arrows: Lines in a circuit diagram represent the conductive paths or wires that connect the components. Arrows indicate the direction of current flow within the circuit.

3. Labels and Values: Circuit diagrams often include labels and values associated with the components, such as resistance values for resistors or voltage ratings for capacitors. These labels provide additional information and aid in circuit analysis.

Types of Circuit Diagrams:
1. Schematic Diagrams: Schematic diagrams are the most common type of circuit diagram. They represent the electrical connections and components of a circuit

without necessarily depicting the physical layout. Schematic diagrams focus on the electrical functionality and relationships within the circuit.

2. Wiring Diagrams: Wiring diagrams, as the name implies, show the physical layout and interconnections of components in a circuit. They provide a detailed representation of how the components are physically connected and can be useful for installation or troubleshooting purposes.

3. Block Diagrams: Block diagrams represent a high-level overview of a system or circuit, focusing on the functional blocks or major components. They are often used to illustrate the overall structure and relationships between subsystems in complex systems.

Benefits of Circuit Diagrams:

1. Visualization: Circuit diagrams provide a visual representation of a circuit, allowing for a clear understanding of how the components are connected and how the circuit functions.

2. Design and Analysis: Circuit diagrams are essential for designing and analyzing circuits. They help engineers and designers plan the layout, choose appropriate components, and ensure proper functionality before building the physical circuit.

3. Troubleshooting: Circuit diagrams aid in troubleshooting electrical systems. By following the diagram and comparing it to the actual circuit, technicians can identify faulty components, incorrect connections, or other issues.

4. Documentation: Circuit diagrams serve as documentation for circuits. They allow for easy sharing of circuit designs, replication of circuits, and future reference.

It's important to note that circuit diagrams may vary in complexity depending on the circuit's purpose and the intended audience. While simple circuits can be represented with basic symbols and connections, complex circuits may require more detailed diagrams with additional annotations and labels.

Overall, circuit diagrams are an essential tool in electrical and electronic engineering, providing a standardized and concise representation of circuits for analysis, design, and communication purposes.

Chapter 2: Passive Components

A. Resistors: Types, Color Coding, And Applications

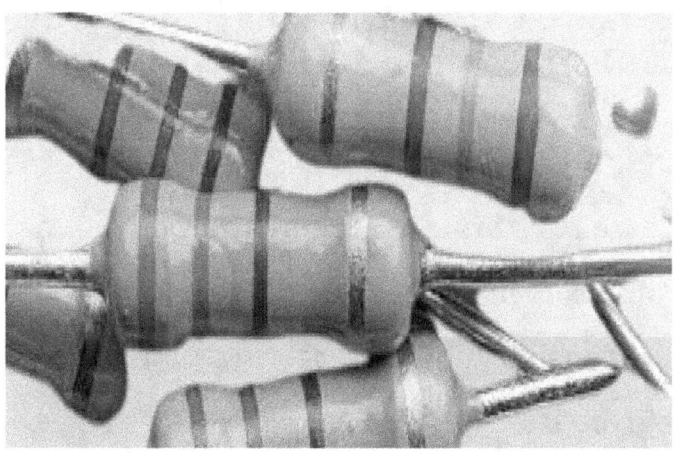

Resistors are passive electronic components that restrict the flow of electric current in a circuit. They are widely used in various electronic applications to control current, divide voltage, and limit power. Here's an overview of resistors, including their types, color coding, and applications:

Types of Resistors:

1. Carbon Composition Resistors:

These resistors are made of a carbon-based composite material. They are relatively inexpensive but have larger size tolerances and are less stable compared to other types. They are frequently employed in applications that operate at low frequencies.

2. Film Resistors:

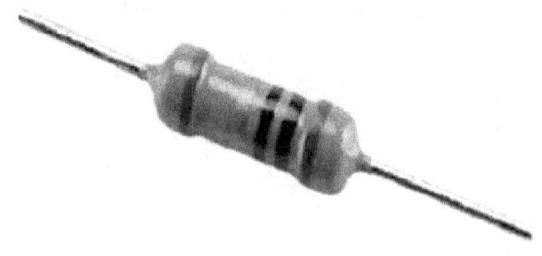

Film resistors are made by depositing a resistive material onto an insulating substrate. They come in two types:

  - Carbon Film Resistors: These resistors have a thin carbon-based film. They provide good stability, accuracy, and low noise performance.

  - Metal Film Resistors: Metal film resistors use a thin metal layer (such as tin-oxide or nickel-chromium) on the substrate. They offer high precision, stability, and low temperature coefficient of resistance (TCR).

3. Wirewound Resistors:

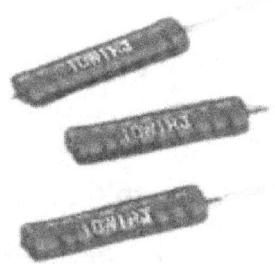

Wirewound resistors are made by winding a resistive wire (usually nichrome) around a ceramic or fiberglass core. They can handle high power and offer high precision and stability. They are commonly used in power applications and high-frequency circuits.

4. **Surface Mount Resistors:**

These resistors are designed for surface mount technology (SMT) and have a small, compact size. They are widely used in modern electronics due to their space-saving nature and suitability for automated assembly processes.

Color Coding of Resistors:

Many through-hole resistors use a color code system to indicate their resistance value, tolerance, and sometimes their temperature coefficient. The color code consists of colored bands painted on the resistor body. Each color represents a specific digit or multiplier value. The standard color code follows the sequence:

- First band: First significant digit
- Second band: Second significant digit
- Third band: Multiplier (number of zeros)
- Fourth band: Tolerance (if present)

Applications of Resistors:

Resistors have numerous applications in electronic circuits, including:

1. Voltage and Current Division: Resistors are used to divide voltage or current in a circuit, providing specific voltage or current levels required by different components.

2. Current Limiting: Resistors are commonly used as current-limiting resistors in LEDs to prevent excessive current flow and protect the LED from damage.

3. Voltage Dropping: Resistors can be used to drop voltage levels in a circuit, for example, in voltage dividers or biasing circuits.

4. Timing and Oscillators: Resistors, in combination with capacitors, are used to set the timing and frequency in timing circuits and oscillators.

5. Signal Conditioning: Resistors are used in signal conditioning circuits to adjust signal levels, impedance matching, and filtering.

6. Temperature Sensing: Some resistors, such as thermistors, exhibit resistance changes with temperature and are used for temperature sensing and compensation.

7. Pulldown and Pullup Resistors: These resistors are used in digital circuits to ensure stable logic levels by providing a known resistance path to either ground (pulldown) or power supply (pullup).

8. Power Dissipation: High-power resistors are used in circuits that require dissipation of significant amounts of power, such as in power supplies and amplifiers.

B. Capacitors: Types, Capacitance, And Charging/Discharging

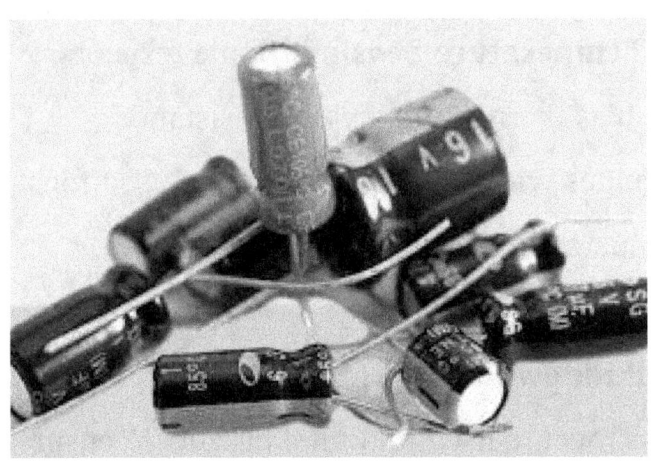

Capacitors are passive electronic components that store and release electrical energy in the form of an electric field. They consist of two conductive plates separated by a dielectric material. Capacitors have various types, are characterized by capacitance values, and play a crucial role in circuits. Let's explore capacitors in more detail:

Types of Capacitors:

1. **Ceramic Capacitors:**

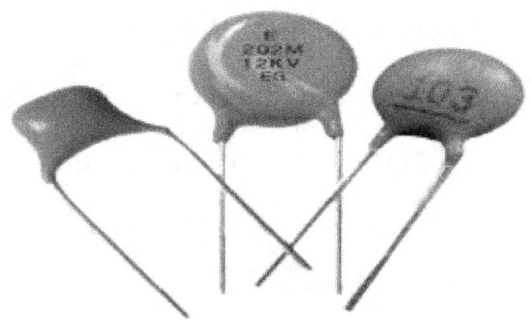

Ceramic capacitors are small, inexpensive, and widely used. They use a ceramic as the dielectric material and come in different types, including class 1 (high accuracy and stability) and class 2 (higher capacitance values).

2. **Electrolytic Capacitors:**

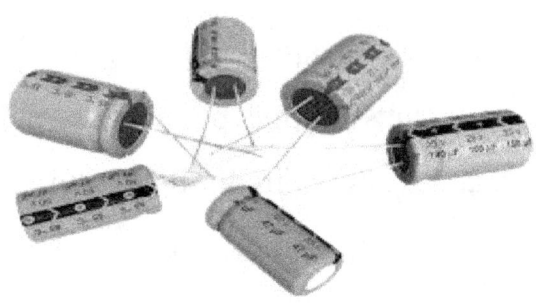

Electrolytic capacitors have a higher capacitance compared to ceramic capacitors. They use an electrolyte-soaked paper as the dielectric material. Electrolytic capacitors are polarized and have a positive and negative terminal. They are commonly used in power supply circuits.

3. **Film Capacitors:**

Film capacitors use a plastic or polymer film as the dielectric material. They have good stability, low losses, and a wide range of capacitance values. Film capacitors are used in various applications, including audio circuits, filters, and timing circuits.

4. Tantalum Capacitors:

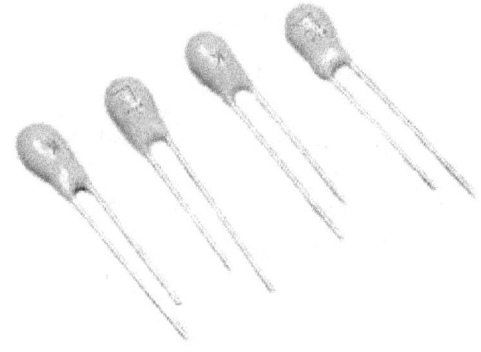

Tantalum capacitors are electrolytic capacitors that use tantalum as the anode material. They offer high capacitance in a small package and are commonly used in portable electronic devices.

5. Aluminum Electrolytic Capacitors:

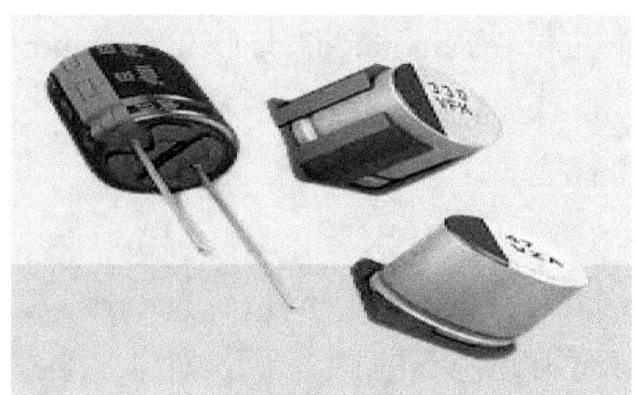

Aluminum electrolytic capacitors are electrolytic capacitors that use aluminum as the anode material. They have high capacitance values and are commonly used in power supply circuits.

Capacitance:

Capacitance is a metric that quantifies the capability of a capacitor to accumulate and store electric charge. It is measured in farads (F), although capacitance values for practical capacitors are typically in microfarads (µF), nanofarads (nF), or picofarads (pF). The capacitance value determines the amount of charge a capacitor can store per volt applied across its terminals.

Capacitance depends on three factors:

1. Area of the Plates: The larger the surface area of the plates, the higher the capacitance.

2. Distance between the Plates: The closer the plates are, the higher the capacitance.

3. Dielectric Material: The type of dielectric material used affects the capacitance. Different materials have different permittivity values, which determine the capacitance.

Charging and Discharging of Capacitors: When a capacitor is connected to a voltage source, such as a battery, it charges up and stores electric energy. The process of charging a capacitor involves the flow of current from the source to the capacitor until it reaches its maximum charge. The time it takes for a capacitor to charge depends on the capacitance value and the resistance in the circuit.

The charging process follows an exponential curve and can be described by the equation:

$$V(t) = V_0 * (1 - e^{(-t / RC)})$$

Where:
- $V(t)$ is the voltage across the capacitor at time t.
- V_0 is the final voltage (source voltage).
- t is the time elapsed.
- R is the resistance in the circuit.
- C is the capacitance of the capacitor.

When a charged capacitor is disconnected from the voltage source, it begins to discharge. The discharge process also follows an exponential curve and can be described by a similar equation.

Charging and discharging of capacitors have various applications, including energy storage, timing circuits, filters, smoothing voltage fluctuations, and providing short-term power during power interruptions.

C. Inductors: Basics And Inductance

Inductors are passive electronic components that store energy in the form of a magnetic field when current flows through them. They are made of a coil of wire wound around a core material, which can be air, iron, or a magnetic material. Inductors have various applications in electronics, including energy storage, filtering, and impedance matching. Let's explore the basics of inductors and the concept of inductance:

Basics of Inductors:

1. Construction: An inductor consists of a coil of wire wound around a core. The core material affects the inductor's performance and can be chosen based on factors such as desired inductance, frequency range, and current handling capabilities.

2. Inductance: Inductance is the fundamental property of an inductor. It represents its ability to store energy in the form of a magnetic field when a current flows through it. Inductance is measured in henries (H), although millihenries (mH) and microhenries (μH) are more commonly used.

3. Magnetic Field: When current flows through an inductor, a magnetic field is generated around the coil. This magnetic field stores energy.

4. Self-Inductance: Inductance can be thought of as self-inductance because it is related to the inductor's own magnetic field generated by the current flowing through it.

Inductance (L):

Inductance is a measure of the amount of magnetic flux generated by an inductor for a given current. It depends on several factors:

1. Number of Turns: Increasing the number of turns in the coil increases the inductance.

2. Cross-Sectional Area: Increasing the cross-sectional area of the coil increases the inductance.

3. Core Material: The type of core material affects the inductance. Magnetic materials increase the inductance significantly compared to air cores.

4. Coil Length: Increasing the length of the coil decreases the inductance.

The unit of inductance, the henry (H), is quite large for most practical applications. Therefore, millihenries (mH) and microhenries (µH) are commonly used.

Applications of Inductors:
Inductors have various applications in electronics, including:

1. Energy Storage: Inductors store energy in their magnetic fields and release it when the current changes. They are used in energy storage applications such as inductors in power supplies and transformers.

2. Filters: Inductors, along with capacitors, form filters for specific frequency ranges. They can pass or block certain frequencies, making them useful in audio, radio, and power filtering applications.

3. Inductive Loads: Inductive loads, such as motors and solenoids, rely on inductors to convert electrical energy into mechanical energy.

4. Inductive Kickback Protection: Inductors can protect circuits from voltage spikes or "kickback" that occurs when a current flowing through an inductor is abruptly interrupted. They absorb and dissipate the energy to prevent damage to other components.

5. Impedance Matching: Inductors are used in impedance matching circuits to ensure maximum power transfer between source and load.

6. Oscillators and Timing Circuits: Inductors, along with capacitors, are used in oscillators and timing circuits to generate precise frequencies and time intervals.

D. Understanding Passive Component Values

Understanding passive component values is crucial for designing and analyzing electronic circuits. Passive components encompass resistors, capacitors, and inductors. Here's an overview of how to interpret and work with the values of these components:

Resistors:
The value of a resistor is typically indicated using a color code system or by specifying the resistance value in ohms (Ω) directly. The color code system is commonly used for through-hole resistors and consists of colored bands painted on the resistor body. Each color represents a specific digit or multiplier value. The standard color code sequence is as follows:

- First band: First significant digit
- Second band: Second significant digit
- Third band: Multiplier (number of zeros)
- Fourth band: Tolerance (if present)

For example, a resistor with bands of brown-black-red-gold represents a resistance value of 1-0-2 with a multiplier of 100 (1000 Ω or 1 kΩ) and a tolerance of ±5%.

Capacitors:

The capacitance value of a capacitor is typically indicated using the metric prefixes micro (μ), nano (n), and pico (p) farads. For example, a capacitor labeled as 10 μF has a capacitance of 10 microfarads, 10 nF represents 10 nanofarads, and 10 pF represents 10 picofarads.

In addition to the capacitance value, capacitors may also have voltage ratings and tolerances specified. The voltage rating indicates the maximum voltage that the capacitor can safely handle. The tolerance specifies the allowable deviation from the stated capacitance value.

Inductors:

The value of an inductor, representing its inductance, is typically specified in henries (H), millihenries (mH), or microhenries (μH). Inductance values are generally larger compared to resistors and capacitors. For example, an inductor labeled as 10 mH has an inductance of 10 millihenries.

Inductors may also have current ratings and tolerances specified. The current rating indicates the maximum current that the inductor can handle without significant performance degradation or damage. The tolerance specifies the allowable deviation from the stated inductance value.

Chapter 3: Active Components

A. Introduction To Semiconductors

Semiconductors are a fundamental class of materials that have properties between those of conductors and insulators. They are crucial components in electronic devices and form the basis of modern electronics. Semiconductors have unique electrical characteristics that allow them to control the flow of electric current. Here's an introduction to semiconductors:

Structure:

Semiconductors are typically crystalline solids composed of atoms arranged in a regular pattern. The most common semiconductor material is silicon (Si), but other materials like germanium (Ge) and compound semiconductors (such as gallium arsenide, GaAs) are also used. The arrangement of atoms in a semiconductor creates a band structure, consisting of valence bands and conduction bands.

Valence Band and Conduction Band:

The valence band is the energy band in which valence electrons (outermost electrons of an atom) reside. In an undisturbed semiconductor, this band is completely filled with electrons. The conduction band, on the other hand, is the energy band above the valence band that is vacant or partially filled with electrons.

Electrons in the conduction band are free to move and contribute to the electric current flow.

Band Gap:

The band gap is the energy difference (gap) between the valence band as well as the conduction band. It determines the electrical behavior of a semiconductor.

Semiconductors can be categorized into two types, depending on their band gaps:

1. Intrinsic Semiconductors: Intrinsic semiconductors have a band gap that is neither too large nor too small. At room temperature, they have a small number of electrons in the conduction band, allowing them to conduct electricity to some extent. Intrinsic semiconductors, such as silicon and germanium, serve as examples in this context.

2. Extrinsic Semiconductors: Extrinsic semiconductors are created by introducing impurities into the crystal lattice of an intrinsic semiconductor. This process is called doping. Doping allows control over the electrical conductivity of the semiconductor. There are two commonly utilized types of doping:

  - N-Type Semiconductor: Doping with impurities that introduce extra electrons (called donor impurities) creates an excess of electrons in the conduction band, resulting in an n-type semiconductor.

  - P-Type Semiconductor: Doping with impurities that create "holes" (electron vacancies) in the valence band (called acceptor impurities) creates a deficit of electrons, leading to a p-type semiconductor.

Electrical Behavior and Applications: The electrical behavior of semiconductors can be controlled by manipulating their doping levels and applying electric fields. Some important semiconductor devices include:

1. Diodes: Diodes are two-terminal devices that allow current to flow in one direction while blocking it in the opposite direction. They are widely used in rectifiers, voltage regulators, and signal demodulation.

2. Transistors: Transistors are three-terminal devices that amplify and switch electronic signals. They are the building blocks of digital logic circuits and amplifiers.

3. Integrated Circuits (ICs): Integrated circuits consist of numerous interconnected transistors and other components fabricated on a single semiconductor substrate. They form the basis of modern electronic devices, including microprocessors, memory chips, and communication systems.

4. Light-Emitting Diodes (LEDs): LEDs are semiconductor devices that emit light when current flows through them. They are used in various lighting applications, displays, and indicators.

Semiconductors have revolutionized the field of electronics by enabling the development of compact, high-performance electronic devices. The ability to control and manipulate the flow of electrons in semiconductors has led to advancements in computing, telecommunications, energy, and many other industries.

B. Diodes: Types And Applications

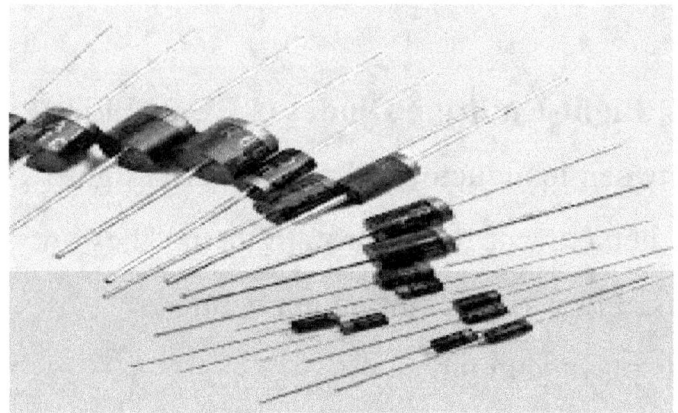

Diodes are two-terminal electronic devices that allow current to flow in one direction while blocking it in the opposite direction. They are crucial components in various electronic circuits and have a wide range of applications. Let's explore the types of diodes and their applications:

1. Rectifier Diodes:

Rectifier diodes are the most common type of diodes and are used to convert alternating current (AC) into direct current (DC). They allow current to flow in the forward direction (from anode to cathode) while blocking it in the reverse direction. Rectifier diodes are widely used in power supplies, battery chargers, and voltage regulators.

2. Zener Diodes:

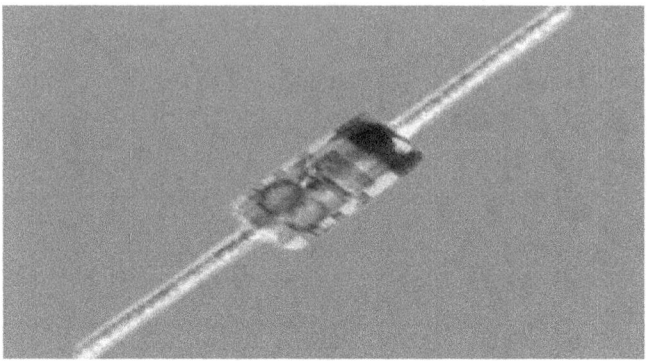

Zener diodes are specially designed to operate in reverse breakdown voltage regions. They have a specific breakdown voltage known as the Zener voltage. When the voltage across the diode reaches the Zener voltage, it conducts in the reverse direction, allowing precise voltage regulation and voltage reference. Zener diodes are commonly used in voltage regulation, voltage clamping, and surge protection applications.

3. Light-Emitting Diodes (LEDs):

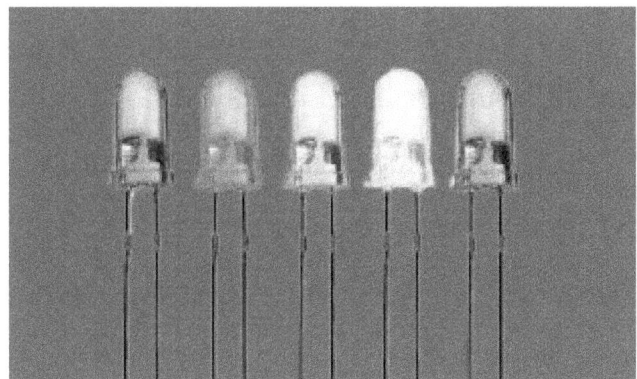

LEDs are semiconductor devices that emit light when current flows through them. They are widely used in lighting applications, display panels, indicators, and signage. LEDs offer energy efficiency, long life, and are available in various colors.

4. Schottky Diodes:

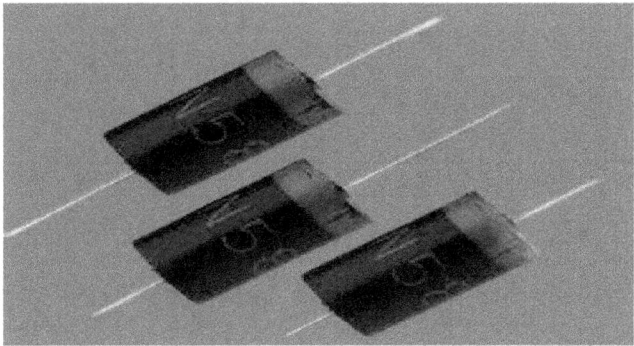

Schottky diodes are characterized by their low forward voltage drop and fast switching speed. They are made of a metal-semiconductor junction and are commonly used in high-frequency and high-speed applications, such as in power supplies, RF circuits, and rectifiers.

5. Varactor Diodes:

Varactor diodes, also known as voltage variable capacitors, are used for voltage-controlled tuning, modulation, and frequency control. Their capacitance varies with the applied voltage, allowing them to function as voltage-dependent capacitors.

Varactor diodes are commonly used in tunable filters, voltage-controlled oscillators, and frequency synthesizers.

6. PIN Diodes:

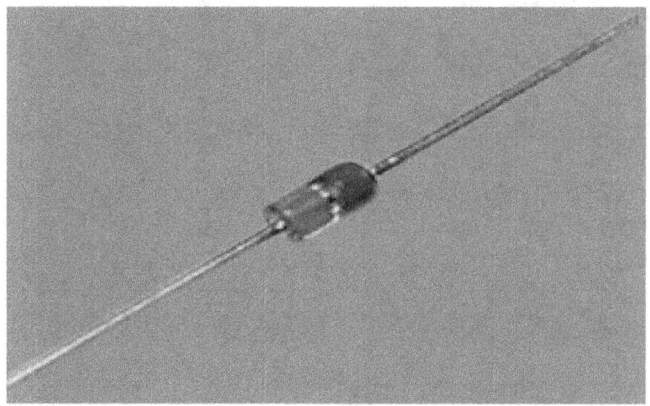

PIN diodes are constructed with an intrinsic (undoped) layer sandwiched between a p-type and an n-type layer. They exhibit a wide depletion region, making them suitable for applications requiring low capacitance and fast switching speeds. PIN diodes are used in RF switches, attenuators, and high-frequency applications.

7. Photodiodes:

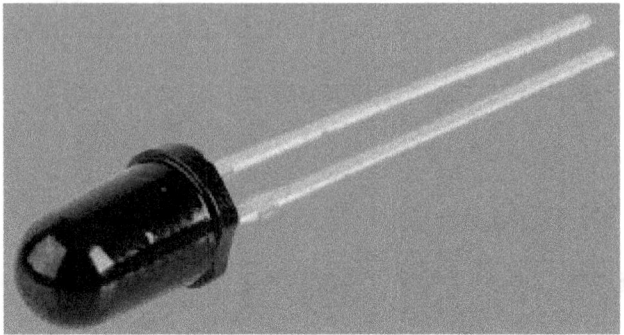

Photodiodes are designed to detect light and convert it into an electric current. They are used in various applications, including optical communication systems, light sensors, and solar cells.

The aforementioned examples merely represent a subset of diode types and their respective applications. Diodes play a critical role in controlling current flow, converting AC to DC, regulating voltages, protecting circuits, and enabling various electronic functions in a wide range of electronic devices and systems.

C. Transistors: Bipolar Junction Transistors (BJTs) and Field-Effect Transistors (FETs)

Transistors are three-terminal electronic devices that amplify and switch electronic signals. They are key components in modern electronics and are categorized into two main types: Bipolar Junction Transistors (BJTs) and Field-Effect Transistors (FETs). Let's explore each type:

1. Bipolar Junction Transistors (BJTs):

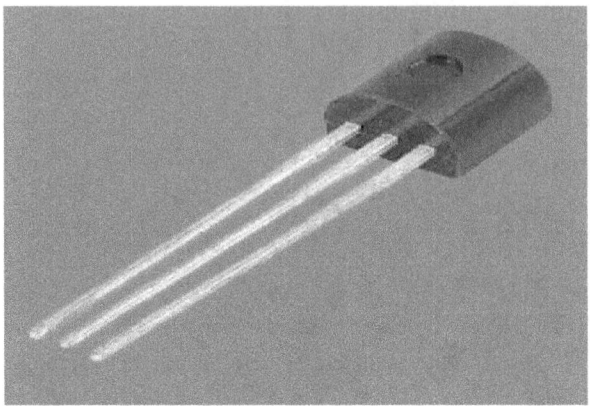

BJTs are composed of three layers of semiconductor material: the emitter, base, and collector. They are classified into two types: NPN (Negative-Positive-Negative) and PNP (Positive-Negative-Positive).

- NPN Transistor: In an NPN transistor, the emitter is made of N-type material, while the base and collector are made of P-type material. The current flows from the emitter to the base, controlling the larger current flowing from the collector to the emitter.

NPN transistors are commonly used in amplification and switching applications.

- PNP Transistor: In a PNP transistor, the emitter is made of P-type material, while the base and collector are made of N-type material. The current flows from the base to the emitter, controlling the larger current flowing from the emitter to the collector. PNP transistors are also used in amplification and switching circuits.

BJTs operate by controlling the current flowing between the collector and emitter terminals based on the current or voltage applied to the base terminal. They have high current gain (hFE or β) and can provide amplification. BJTs are used in various applications, including amplifiers, switching circuits, oscillators, and digital logic circuits.

2. Field-Effect Transistors (FETs):

FETs are three-terminal devices that rely on the control of an electric field to modulate the conductivity of a semiconductor channel. They are categorized into two main types: MOSFETs (Metal-Oxide-Semiconductor Field-Effect Transistors) and JFETs (Junction Field-Effect Transistors).

- MOSFETs: MOSFETs have a metal-oxide-insulator layer between the gate and the channel, which is usually made of silicon dioxide (SiO_2). They are further classified into two types: enhancement-mode and depletion-mode MOSFETs. MOSFETs are the most widely used type of FET and are employed in a wide range of applications, including digital circuits, power amplifiers, voltage regulators, and integrated circuits.

- JFETs: JFETs have a junction between the gate and the channel, which is formed by the reverse-biased PN junction. JFETs are available in two types: N-channel and P-channel. Field-effect transistors (FETs) function by regulating the width of the conducting channel between the source and drain terminals through the application of voltage to the gate terminal. JFETs are commonly used in low-noise amplifiers, switches, and voltage-controlled resistors.

FETs offer high input impedance, low power consumption, and can operate at high frequencies. They are suitable for applications that require high impedance inputs, low noise, and fast switching speeds.

Both BJTs and FETs are essential in modern electronics, and their selection depends on specific circuit requirements such as voltage levels, current levels, speed, power dissipation, and noise considerations. Understanding the characteristics and applications of BJTs and FETs is important for designing and analyzing electronic circuits effectively.

D. Integrated Circuits (ICs) And Their Uses

Integrated Circuits (ICs), also known as microchips or chips, are miniaturized electronic circuits that contain a large number of electronic components, such as transistors, resistors, capacitors, and diodes, fabricated on a single semiconductor substrate. ICs have revolutionized the field of electronics by enabling the integration of complex circuitry into small packages. Here are some key points about ICs and their uses:

1. Types of ICs:
- **Digital ICs:**

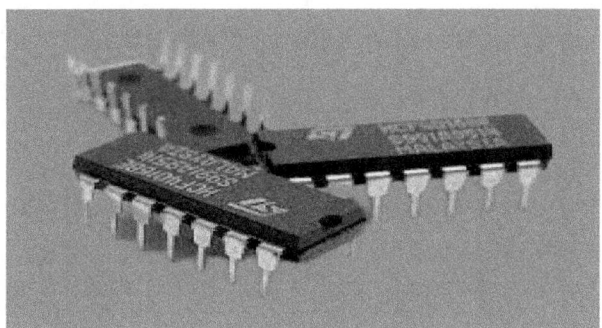

These ICs are designed to process and manipulate digital signals. They form the basis of digital logic circuits used in computers, microcontrollers, memory chips, and digital signal processors (DSPs).

- **Analog ICs:**

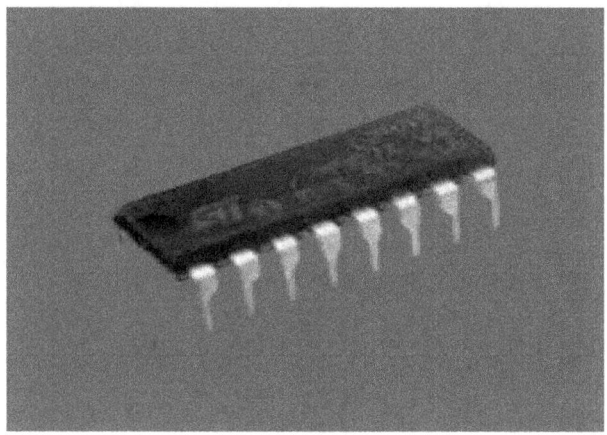

Analog ICs are used to process continuous analog signals. They include operational amplifiers (op-amps), voltage regulators, audio amplifiers, and analog-to-digital converters (ADCs) and digital-to-analog converters (DACs).

- Mixed-Signal ICs:

Mixed-signal ICs combine both digital and analog circuitry on a single chip. They are used in applications that require both analog and digital processing, such as data acquisition systems, sensors, and communication systems.

2. Functions and Applications of ICs:
- Microprocessors and Microcontrollers:
ICs like microprocessors and microcontrollers are the heart of computer systems, embedded systems, and consumer electronics. They perform tasks like data processing, control, and communication.

- **Memory Chips:** ICs such as RAM (Random Access Memory) and ROM (Read-Only Memory) chips are used for data storage in computers, smartphones, and other devices.

- **Power Management ICs:** These ICs regulate and control power distribution and conversion in electronic devices. They are used in power supplies, battery charging systems, and energy management applications.

- **Communication ICs:** ICs for communication systems include radio frequency (RF) ICs, baseband processors, and modem chips used in wireless communication, cellular networks, Wi-Fi, and Bluetooth devices.

- **Sensor ICs:** These ICs interface with various sensors, such as temperature sensors, pressure sensors, motion sensors, and environmental sensors, to convert physical signals into electrical signals for processing.

- **Display Driver ICs:** These ICs control the operation and drive the pixels of display panels, such as LCDs (Liquid Crystal Displays) and OLEDs (Organic Light-Emitting Diodes).

- **Audio and Video ICs:** ICs for audio and video applications include audio amplifiers, video processors, codecs, and video encoders/decoders used in multimedia systems, televisions, and audio/video equipment.

- **Automotive ICs:** ICs designed for automotive applications include engine control units (ECUs), sensors, motor control ICs, and infotainment systems used in vehicles.

- **Industrial and IoT ICs:** ICs for industrial applications include motor control ICs, process control ICs, and interface ICs used in factory automation, robotics, and Internet of Things (IoT) devices.

Integrated Circuits have made electronic devices smaller, more efficient, and more reliable. They have enabled the development of advanced technologies and have become an integral part of our daily lives, powering everything from smartphones and computers to cars and medical equipment. The versatility and widespread use of ICs continue to drive innovation and advancements in various fields of technology.

Chapter 4: Circuit Analysis

A. Series And Parallel Circuits

Series and parallel circuits are two basic configurations for connecting electrical components in a circuit. Understanding the differences between series and parallel connections is crucial in analyzing and designing electronic circuits. Here's an overview of series and parallel circuits:

1. Series Circuits:

In a series circuit, components are connected sequentially, one after another, creating a single pathway for current flow. The main characteristics of a series circuit are as follows:

- **Current:** The same current flows through each component in a series circuit. The total current in the circuit is the combined value of the individual currents flowing through each component.

- Voltage: The overall voltage across the circuit is the accumulation of the individual voltage drops across each component. The voltage is divided among the components in proportion to their resistance or impedance.

- **Resistance:** In a series circuit, the total resistance is the summation of the resistances of each component. The total resistance of a circuit rises proportionally with the addition of more resistors in series configuration.

- **Brightness:** In a series circuit with identical light bulbs, the brightness of each bulb decreases as more bulbs are added because the total resistance increases.

2. Parallel Circuits:

In a parallel circuit, components are connected across each other, providing multiple paths for current flow. The main characteristics of a parallel circuit are as follows:

- **Current:** The total current in a parallel circuit is divided among the parallel branches, and the current flowing through each component is determined by its individual resistance or impedance.

- **Voltage:** The voltage across each component in a parallel circuit is the same. In a circuit, the total voltage across all components is equivalent to the sum of the individual voltages across each component within the circuit.

- **Resistance:** In a parallel circuit, the total resistance demonstrates an inverse relationship with the summation of the reciprocals of the individual resistances, signifying that as the sum of the reciprocals increases, the total resistance decreases. Adding more resistors in parallel decreases the total resistance.

- **Brightness:** In a parallel circuit with identical light bulbs, each bulb has the same brightness because they all receive the same voltage.

3. Combination Circuits:

In practice, most circuits are a combination of series and parallel connections. Complex circuits can be broken down into simpler series and parallel sections for analysis.

- Mixed Series-Parallel Circuits: These circuits have components connected in both series and parallel configurations. Analyzing such circuits often involves breaking them down into simpler series and parallel sections.

Understanding the characteristics of series and parallel circuits helps in determining voltage drops, current flows, and overall circuit behavior. It also aids in calculating the equivalent resistance or impedance of a circuit, which simplifies circuit analysis and design.

B. Kirchhoff's Laws And Their Application

Kirchhoff's laws, named after German physicist Gustav Kirchhoff, are fundamental principles used in the analysis of electrical circuits. They provide a set of rules for the conservation of charge and energy in a circuit. Let's explore Kirchhoff's two laws and their applications:

1. Kirchhoff's Voltage Law (KVL):
Kirchhoff's Voltage Law states that the sum of the voltage drops (or rises) around any closed loop in a circuit is equal to the sum of the voltage sources in that loop. Mathematically, it can be stated as:

$\Sigma V = 0$

Where ΣV is the sum of all the voltage drops and rises around a closed loop.

Applications of KVL:
- Determining unknown voltages: KVL is used to calculate the voltage across resistors, capacitors, and other circuit elements in a closed loop.

- **Analyzing complex circuits:** KVL is especially useful in analyzing circuits with multiple loops, where it helps in writing and solving simultaneous equations to determine unknown voltages or currents.

- Verifying circuit calculations: KVL is also used to verify calculations in circuit analysis by ensuring that the sum of the voltage drops equals the sum of the voltage sources.

2. Kirchhoff's Current Law (KCL):
Kirchhoff's Current Law states that the sum of currents entering a node (or junction) in a

circuit is equal to the sum of currents leaving that node. In other words, the total current flowing into a node is equal to the total current flowing out of the node. Mathematically, it can be stated as:

$$\Sigma I = 0$$

Where ΣI is the sum of all the currents entering or leaving a node.

Applications of KCL:
- Current division: KCL is used to determine the distribution of current among parallel branches in a circuit.
- **Nodal analysis:** KCL forms the basis of nodal analysis, a technique used to analyze complex circuits by writing and solving simultaneous equations based on current conservation at each node.

- Checking circuit connections: KCL is useful in verifying circuit connections by ensuring that the sum of currents entering a node equals the sum of currents leaving that node.

Kirchhoff's laws are essential tools for circuit analysis, providing a systematic and mathematically rigorous approach to solving complex circuits. They are widely used in electrical engineering and are fundamental in designing and troubleshooting electronic circuits. By applying Kirchhoff's laws, engineers can determine voltages, currents, and power dissipation in circuits, enabling the proper functioning and optimization of electronic systems.

C. Voltage And Current Division

Voltage Division and Current Division are two fundamental principles used to determine how voltage and current are distributed in a circuit with multiple resistors or components connected in series or parallel. Let's explore each concept:

1. Voltage Division:

Voltage Division refers to the distribution of voltage across multiple resistors connected in series. According to the Voltage Division Rule, the voltage across each resistor in a series circuit is proportional to its resistance.

The formula for voltage division is as follows:
V1 = (R1 / (R1 + R2 + ... + Rn)) * V

Where V1 is the voltage across resistor R1, R1, R2, ..., Rn are the resistances in the series circuit, V is the total voltage applied to the circuit.

Applications of Voltage Division:
- Determining voltage across specific resistors: By using the voltage division formula, you can calculate the voltage drop across each resistor in a series circuit.
- Level shifting: Voltage division is often used to shift the voltage level from a higher value to a lower value in electronic circuits.

2. Current Division:
Current Division refers to the distribution of current among parallel branches in a circuit. According to the Current Division Rule, the current flowing through each branch of a parallel circuit is inversely proportional to the resistance of that branch.

The formula for current division is as follows:

$I_1 = (R_{tot} / R_1) * I$

Where I_1 is the current flowing through branch R_1, R_{tot} is the total resistance of the parallel circuit, I is the total current entering the parallel circuit.

Applications of Current Division:

- Determining current through specific branches: Using the current division formula, you can calculate the current flowing through each branch of a parallel circuit.
- Current sharing: Current division is often used to ensure that different components or loads in parallel receive the desired amount of current.

Both voltage division and current division are valuable tools in circuit analysis and design. They allow engineers to determine the voltage drops and current flows in series and parallel circuits, which is essential for understanding the behavior of electronic systems and optimizing their performance.

D. Introduction To Breadboarding

Breadboarding is a fundamental technique used in electronics for prototyping and testing circuits. It involves using a breadboard, also known as a prototyping board or solderless breadboard, to build temporary circuits without the need for soldering or permanent connections. Breadboarding allows for easy experimentation, modification, and troubleshooting of electronic circuits. Here's an introduction to breadboarding:

1. Breadboard Overview:

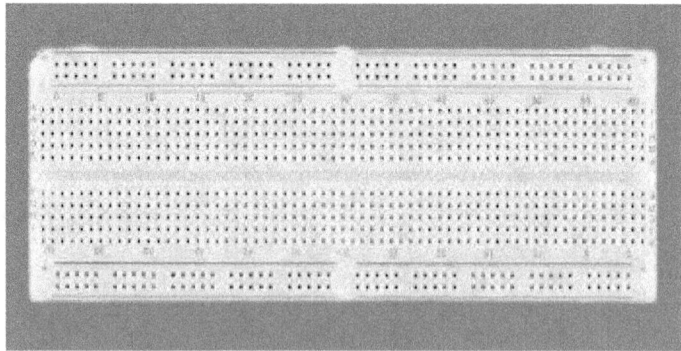

A breadboard is typically a rectangular plastic board with a grid of holes and metal clips inside. The board contains rows and columns of interconnected holes that allow components to be inserted and connected together. The holes in the breadboard are arranged in a specific pattern, usually following a standard layout.

2. Terminal Strips:

The breadboard usually consists of two terminal strips running along the longer sides of the board. Each terminal strip consists of multiple rows of interconnected holes. The holes within each row are electrically connected, allowing for the placement of components and connection of wires.

3. Power Rails:

The breadboard typically has two horizontal strips known as power rails. These power rails are used for supplying power to the circuit. The power rails are typically color-coded, with the red rail (+) representing the positive supply voltage and the blue or black rail (-) representing the ground or negative supply voltage.

4. Component Placement:

Electronic components, such as resistors, capacitors, integrated circuits (ICs), and jumper wires, can be inserted into the holes on the breadboard. Components with leads (legs) are inserted into the appropriate holes to establish connections. The holes within each row and column are electrically connected, allowing for the components to be connected together without the need for soldering.

5. Interconnections:

Components are connected by using jumper wires, which are flexible wires with pins or connectors on each end. Jumper wires are used to bridge the gaps between components, connecting them as desired to create the circuit connections. These wires are inserted into the holes on the breadboard, establishing electrical connections between the components.

6. Prototyping and Testing:

Breadboarding allows for quick and temporary prototyping of circuits. It enables the assembly of circuits to test their functionality, experiment with different component configurations, and troubleshoot any issues without the need for soldering. If modifications are required, components can be easily removed or repositioned on the breadboard.

7. Limitations:

While breadboarding is a convenient prototyping method, it does have its limitations. Breadboards are primarily designed for low-frequency and low-power circuits. High-frequency signals, sensitive analog circuits, and high-current applications may require more specialized techniques and PCB (Printed Circuit Board) design.

Chapter 5: Basic Electronic Devices And Applications

A. Light-Emitting Diodes (LEDs)

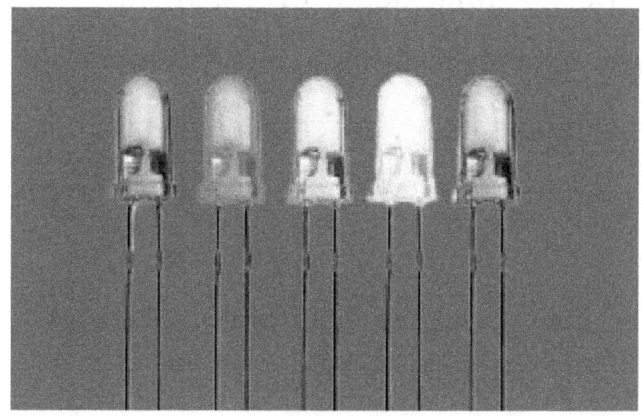

Light-Emitting Diodes (LEDs) are semiconductor devices that emit light when an electric current passes through them. LEDs are widely used in various applications, ranging from indicator lights to displays and lighting systems. Here's an overview of LEDs:

1. Working Principle:

LEDs are based on the principle of electroluminescence, which is the emission of light from a material when it is electrically stimulated. LEDs are made of a semiconductor material, typically composed of a combination of different elements from the periodic table. When a forward voltage is applied to the LED, electrons and holes recombine in the semiconductor, releasing energy in the form of photons (light).

2. Advantages of LEDs:

LEDs offer several advantages over traditional light sources, such as incandescent bulbs and fluorescent lights. Some key advantages include:
- Energy Efficiency: LEDs are highly energy-efficient, converting a significant portion of electrical energy into light, minimizing wasted energy as heat.

- Long Lifespan: LEDs have a much longer lifespan compared to traditional light sources, often lasting tens of thousands of hours.
- Durability: LEDs are solid-state devices with no moving parts, making them more robust and resistant to shock and vibration.
- Compact Size: LEDs are compact and can be designed in various shapes and sizes, allowing for flexible integration into different applications.
- Instantaneous Response: LEDs light up instantly with no warm-up time, providing immediate illumination.

3. Types of LEDs:
- **Single-Color LEDs:** These LEDs emit light in a single color, such as red, green, blue, yellow, or white. White LEDs are often created by combining different-color LEDs or by using phosphor coatings.

- **RGB LEDs:** RGB (Red, Green, Blue) LEDs contain three individual LEDs within a single package, enabling the emission of different colors by adjusting the intensity of each color.

- **High-Power LEDs:** High-power LEDs are designed to handle higher currents and deliver greater light output. They find widespread application in the field of lighting, commonly utilized in various lighting systems and applications.

- **Organic LEDs (OLEDs):** OLEDs are a type of LED that uses organic compounds as the light-emitting material. They offer flexibility, high contrast ratios, and wide viewing angles, making them suitable for display applications.

4. Applications of LEDs:

- **Lighting:** LEDs are extensively used for general lighting purposes, including residential, commercial, and outdoor lighting applications. They offer energy efficiency, long lifespan, and the ability to produce different colors.

- **Displays:** LEDs are used in various display applications, including digital signs, traffic lights, alphanumeric displays, and large video screens.

- **Backlighting:** LEDs are commonly used as backlighting sources in LCD (Liquid Crystal Display) panels for televisions, computer monitors, and smartphones.

- **Indicator Lights:** LEDs are widely used as indicator lights in electronic devices and appliances, providing visual feedback for power status, operation, and warnings.

- **Automotive Lighting:** LEDs are used in headlights, taillights, brake lights, and interior lighting in automotive applications due to their energy efficiency and durability.

- **Decorative Lighting:** LEDs are popular for decorative lighting, such as in architectural lighting, holiday lighting, and decorative fixtures.

B. Switches And Push Buttons

Switches and push buttons are essential components in electronics that allow for the control and manipulation of electrical circuits. They are used to open or close the circuit, interrupt or allow the flow of current, and control the operation of various devices. Here's an overview of switches and push buttons:

1. Switches:

Switches are mechanical devices used to open or close a circuit. They have two or more states: on (closed) or off (open). When the switch is closed, it completes the circuit, allowing current to flow, whereas an open switch breaks the circuit, interrupting the current flow. Some common types of switches include:

- **Toggle Switches:** Toggle switches have a lever or toggle that can be flipped up or down to open or close the circuit. They are commonly used in various applications, such as controlling power to devices or selecting different modes.

- **Slide Switches:** Slide switches have a small slider that moves back and forth to open or close the circuit. They are often used in compact devices or settings where a toggle switch may not be suitable.

- Rocker Switches: Rocker switches have a pivoting lever that rocks back and forth to open or close the circuit. They are commonly found in household appliances, power tools, and automotive applications.

- **Rotary Switches:** Rotary switches have a rotating knob or lever that can be turned to select different positions, each corresponding to a different circuit connection. They are often used for device settings, such as volume control or mode selection.

2. Push Buttons:

Push buttons, also known as momentary switches, are switches that are only active as long as they are being pressed. When the button is released, the circuit returns to its original state. Push buttons are commonly used for momentary actions, such as turning on a device or triggering a specific function. They are available in diverse forms and dimensions, encompassing a range of shapes and sizes, such as:

- Tactile Push Buttons: Tactile buttons have a small button cap that provides a tactile feedback when pressed. They are commonly used in consumer electronics, keyboards, and control panels.

- **Micro Push Buttons:** Micro buttons are compact and often surface-mounted switches used in small electronic devices and printed circuit boards.
- Illuminated Push Buttons: These push buttons have built-in LED indicators that light up when the button is pressed, providing visual feedback.

3. Applications:

Switches and push buttons are used in various applications, including:

- Power control: Switches are used to turn devices on or off, controlling the power supply.
- **Circuit selection:** Switches are used to select different circuits or modes, such as in audio/video equipment or electronic instruments.

- **Control and automation:** Switches and push buttons are used in control systems to trigger specific actions or functions.
- User interfaces: Push buttons are commonly used in user interfaces for input or triggering specific actions in electronic devices.

It's important to consider the electrical ratings and specifications of switches and push buttons to ensure they can handle the required voltage, current, and environmental conditions. Additionally, switches and push buttons can be combined with other components, such as resistors, capacitors, or microcontrollers, to create more complex control systems.

C. Buzzer And Speaker Circuits

Buzzer and speaker circuits are used to generate audible sound in electronic devices. They are widely employed in applications such as alarms, notifications, audio playback, and communication systems. Here's an overview of buzzer and speaker circuits:

1. Buzzer Circuit:

A buzzer is a simple sound-producing device that typically operates at a fixed frequency. It consists of a piezoelectric element or an electromagnetic coil that vibrates when an electrical signal is applied. The vibration creates sound waves, producing the desired audible tone.

The basic circuit for driving a buzzer typically includes the following components:

- Buzzer: The buzzer element, either a piezoelectric element or an electromagnetic coil.

- Driver Circuit: The driver circuit provides the necessary current and voltage to the buzzer to make it vibrate and produce sound. It can consist of a transistor, a timer IC, or a microcontroller.

The driver circuit applies an oscillating signal to the buzzer element, causing it to vibrate at the desired frequency. The frequency of the sound produced depends on the properties of the buzzer element.

2. Speaker Circuit:

A speaker is a more complex sound reproduction device capable of producing a wide range of frequencies. It consists of a cone or diaphragm connected to a voice coil, which is placed within a magnetic field.

When an electrical signal is applied to the voice coil, it creates vibrations that produce sound waves.

The circuit for driving a speaker is more involved than a buzzer circuit. Typically, it encompasses the following components:
- Speaker: The speaker element, which consists of a cone or diaphragm connected to a voice coil.
- Amplifier: A dedicated audio amplifier or an audio output stage of an electronic device is used to amplify the audio signal to a level sufficient for driving the speaker.
- Audio Signal Source: The audio signal can come from various sources, such as an audio player, a microphone, or a microcontroller generating sound data.

The audio signal is amplified by the amplifier circuit and then fed to the voice coil of the speaker. The varying current in the voice coil causes the cone or diaphragm to vibrate, reproducing the sound waves.

3. Control and Protection:

Both buzzer and speaker circuits can be controlled and protected using additional components. For example:

- Switches: Switches can be used to turn the sound on or off or to activate specific sound patterns.
- Volume Control: Potentiometers or digital volume control circuits can be incorporated to adjust the volume level of the sound output.
- Protection Circuitry: To protect the buzzer or speaker from overcurrent or voltage spikes, appropriate protection circuitry, such as fuses or voltage regulators, can be implemented.

D. Introduction To Sensors: Light, Temperature, And Motion

Sensors are devices that detect and respond to changes in their environment. They play a crucial role in modern technology by providing input to electronic systems, enabling them to monitor and interact with the world around them. Here's an introduction to three commonly used sensors: light sensors, temperature sensors, and motion sensors.

1. Light Sensors:
Light sensors, also known as photocells or photoresistors, detect the presence or intensity of light in their surroundings. They are responsive to changes in light levels and can be used in various applications, such as automatic lighting control, photography, and light intensity measurement.

A frequently encountered light sensor variant is the photoresistor. It is a passive component that exhibits changes in resistance based on the amount of light falling on its surface. As light intensity increases, the resistance of the photoresistor decreases. This property allows it to be used in light-dependent circuits to control the activation or deactivation of devices.

Another variety of light sensor is the photodiode. It is an active semiconductor device that converts light into an electric current. Photodiodes are often used in applications such as light detection, optical communication, and light meters.

2. Temperature Sensors:
Temperature sensors are used to measure and monitor changes in temperature. They

are essential in applications where temperature regulation, control, or monitoring is required. Temperature sensors find applications in HVAC systems, industrial processes, medical devices, and weather monitoring, among others.

The thermistor is a frequently engaged temperature sensor in various applications. It is a specific type of resistor that displays variations in resistance in response to changes in temperature. Thermistors exhibit a negative temperature coefficient (NTC) or positive temperature coefficient (PTC) depending on their construction. NTC thermistors decrease in resistance as the temperature rises, while PTC thermistors increase in resistance with temperature.

Another widely used temperature sensor is the integrated circuit (IC) temperature sensor. These sensors are typically small and accurate and can provide digital temperature readings directly. They often incorporate additional features like temperature compensation and calibration.

3. Motion Sensors:

Motion sensors detect movement or changes in the surrounding environment. They are commonly used in security systems, automatic doors, occupancy detection, and motion-activated lighting. Various types of motion sensors exist, including:

- Passive Infrared (PIR) Sensors: PIR sensors detect changes in infrared radiation emitted by objects within their field of view. They are commonly used in motion-activated lighting systems and security applications.

- Ultrasonic Sensors: Ultrasonic sensors emit ultrasonic waves and measure the time it takes for the waves to bounce back after hitting an object. By analyzing the reflected waves, they can detect motion and measure distances.

- Microwave Sensors: Microwave sensors emit microwave signals and detect changes in the reflected signals caused by moving objects. They are commonly used in automatic door systems and occupancy detection.

- Accelerometers: Accelerometers detect changes in acceleration and can be used as motion sensors. They are commonly found in smartphones, gaming controllers, and motion-capturing devices.

These are just a few examples of sensors used in electronics. There is a wide range of sensors available for various purposes, including pressure sensors, humidity sensors, proximity sensors, and more. Sensors enable devices to interact with the physical world, making them integral to many applications across industries.

Chapter 6: Power Supplies And Voltage Regulation

A. Types Of Power Supplies: Batteries, Ac Adapters, And Voltage Regulators

Power supplies are devices or systems that provide electrical energy to power electronic circuits and devices. They convert input power from a source, such as batteries or wall outlets, into the required output power for the connected devices. Here are three common types of power supplies: batteries, AC adapters, and voltage regulators.

1. Batteries:

Batteries are portable power sources that convert chemical energy into electrical energy. They come in various sizes and chemistries, such as alkaline, lithium-ion, and lead-acid. Batteries are commonly used in portable devices and applications where mobility is required. Some key features of batteries include:

- Portability: Batteries are self-contained power sources that can be easily transported and used without a constant connection to a power outlet.
- Limited Capacity: Batteries have a finite capacity and need to be recharged or replaced when they are depleted.
- Voltage and Chemistry Variations: Different batteries have different voltage ratings and chemical compositions, which

determine their capacity, energy density, and performance characteristics.

2. AC Adapters:

AC adapters, also known as power adapters or chargers, convert AC (alternating current) voltage from a wall outlet into DC (direct current) voltage suitable for powering electronic devices. AC adapters are commonly used to power laptops, smartphones, home appliances, and various other devices. Some key features of AC adapters include:

- AC-to-DC Conversion: AC adapters convert the higher voltage and alternating current from the wall outlet into the lower voltage and direct current required by the connected device.
- Plug and Voltage Compatibility: AC adapters come in different shapes and sizes

to fit specific devices and regional power outlets. They may have interchangeable plugs or voltage selection switches to accommodate different requirements.

3. Voltage Regulators:

Voltage regulators are devices or circuits that stabilize and regulate the voltage supplied to electronic devices. They ensure that the output voltage remains constant despite fluctuations in the input voltage or changes in the load. Voltage regulators are used in various applications to provide stable and regulated power supply. Some types of voltage regulators include:

- Linear Voltage Regulators: Linear voltage regulators use a series pass transistor to regulate the output voltage. They provide a stable output voltage but are less efficient when the input-output voltage differential is high.

- Switching Voltage Regulators: Switching voltage regulators use a switching transistor and an inductor to regulate the output voltage. They offer higher efficiency and are commonly used in applications where high power conversion efficiency is required.

Voltage regulators can be built into power supply units or implemented as separate components. They ensure that electronic devices receive the appropriate and stable voltage necessary for their proper operation.

B. Voltage Regulation Using Zener Diodes

Zener diodes are widely used for voltage regulation in electronic circuits. They are specifically designed to operate in the reverse breakdown region, where they exhibit a sharp breakdown voltage known as the Zener voltage. By utilizing the Zener effect, Zener diodes can maintain a relatively constant voltage across their terminals, even when the current through them changes. This characteristic makes them ideal for voltage regulation applications. Here's how voltage regulation using Zener diodes works:

1. Zener Diode Characteristics:
Zener diodes are designed to have a specific breakdown voltage, known as the Zener voltage (Vz). When the reverse bias voltage across a Zener diode exceeds the Zener

voltage, the diode starts to conduct heavily in the reverse direction, allowing current to flow through it. This reverse breakdown region is where Zener diodes operate as voltage regulators.

2. Zener Diode as a Voltage Regulator:
To use a Zener diode for voltage regulation, it is typically connected in parallel with the load that requires a stable voltage. The Zener diode is reverse biased, meaning the cathode terminal is connected to a higher voltage than the anode terminal.

The Zener diode regulates the voltage across the load by maintaining a constant voltage drop equal to its Zener voltage (Vz). As long as the input voltage remains higher than the Zener voltage, the Zener diode conducts and keeps the voltage across the load constant. If the input voltage exceeds the Zener voltage, the excess voltage is dropped across the Zener diode, preventing the load from receiving a voltage higher than the Zener voltage.

3. Selection of Zener Diode:

When selecting a Zener diode for voltage regulation, consider the following factors:

- Zener Voltage (Vz): Choose a Zener diode with a voltage rating close to the desired regulated voltage.
- Power Rating: Ensure that the Zener diode can handle the power dissipation based on the current passing through it.
- Reverse Breakdown Voltage (Vbr): Select a Zener diode with a breakdown voltage higher than the maximum expected input voltage to maintain regulation.

4. Current Limiting Resistor:

To protect the Zener diode from excessive current and prevent damage, a current-limiting resistor (Rs) is often connected in series with the diode. The resistor limits the current flowing through the Zener diode and the load. Its value can be calculated using Ohm's law: $Rs = (Vin - Vz) / Iz$, where Vin is the input voltage, Vz is the Zener voltage, and Iz is the Zener diode current.

It's important to note that Zener diodes have certain limitations, such as power dissipation and voltage regulation tolerance. In some cases, additional circuitry, such as an emitter follower or an operational amplifier, may be used to improve the voltage regulation performance.

C. Introduction To Linear And Switching Regulators

Linear and switching regulators are two common types of voltage regulators used in electronic circuits to maintain a stable output voltage. Each type has its own characteristics and is suited for different applications. Here's an introduction to linear and switching regulators:

1. Linear Regulators:

Linear regulators, also known as linear voltage regulators, use a linear control element, typically a pass transistor, to regulate the output voltage. They are simple in design and provide accurate voltage regulation. The key components of a linear regulator include a series pass transistor, a voltage reference, and a feedback loop. Here's how linear regulators work:

- The input voltage is applied to the series pass transistor, which acts as a variable resistor.
- The voltage reference provides a stable reference voltage.
- The feedback loop compares the output voltage with the reference voltage and adjusts the pass transistor to maintain a constant output voltage.

Advantages of Linear Regulators:
- Simplicity: Linear regulators have a simple circuit design and are easy to implement.
- Low output ripple and noise: They provide smooth and low-noise output voltage.
- Good transient response: Linear regulators respond quickly to changes in load and input voltage.

Disadvantages of Linear Regulators:

- Inefficiency: Linear regulators dissipate excess energy as heat, resulting in lower efficiency, especially when the input-output voltage differential is high.

- Limited power handling capability: Linear regulators are suitable for low to moderate power applications due to their power dissipation limitations.

- Voltage drop: Linear regulators have a voltage drop across the pass transistor, which results in wasted energy.

2. Switching Regulators:

Switching regulators, also known as switched-mode power supplies (SMPS), use a switching element, such as a transistor or a MOSFET, to control the output voltage. They operate by rapidly switching the input voltage on and off and then filtering it to obtain the desired output voltage. Here's how switching regulators work:

- The input voltage is switched on and off at a high frequency using a switching element.
- An inductor or transformer stores and transfers energy during the switching cycles.
- The switched voltage is rectified and filtered to obtain the desired output voltage.
- Feedback control adjusts the duty cycle of the switching element to maintain a constant output voltage.

Advantages of Switching Regulators:
- High efficiency: Switching regulators are highly efficient as they minimize power dissipation by switching the input voltage.
- Wide input voltage range: They can operate over a wide range of input voltages.
- Higher power handling capability: Switching regulators are suitable for high-

power applications due to their higher efficiency and reduced heat dissipation.

Disadvantages of Switching Regulators:

- Complexity: Switching regulators have a more complex circuit design compared to linear regulators.

- Higher output ripple and noise: Switching regulators may introduce more noise and ripple in the output voltage compared to linear regulators.

- EMI/EMC considerations: Switching regulators can generate electromagnetic interference (EMI) that requires careful circuit layout and filtering.

Switching regulators are commonly used in applications where high efficiency and power handling are crucial, such as power supplies for computers, industrial equipment, and high-power audio systems. Linear regulators are often used in low-power applications or where low noise and simplicity are important, such as precision instrumentation and audio amplifiers.

The choice between linear and switching regulators depends on the specific requirements of the application, including power requirements, efficiency considerations, size constraints, and cost factors.

Chapter 7: Integrated Circuits And Microcontrollers

A. Introduction To IC Families: TTL, CMOS, And Op-Amps

IC (Integrated Circuit) families refer to different types or families of integrated circuits that are designed with specific characteristics and technologies. Two commonly used IC families are TTL (Transistor-Transistor Logic) and CMOS (Complementary Metal-Oxide-Semiconductor), while Op-Amps (Operational Amplifiers) are a specific type of IC used for analog signal processing. Here's a concise introduction to each:

1. TTL (Transistor-Transistor Logic):
TTL is a popular IC family that uses bipolar junction transistors for its digital logic gates.

It was one of the first widely used IC families and has been in use for several decades. TTL logic gates are known for their high noise immunity, fast switching speeds, and compatibility with different logic families.

TTL ICs operate using a 5-volt power supply and have standardized logic levels: low logic level (0V to 0.8V) and high logic level (2V to 5V). They are commonly used in applications such as microprocessors, digital logic circuits, and interface circuits.

2. CMOS (Complementary Metal-Oxide-Semiconductor):

CMOS is another widely used IC family, known for its low power consumption, high noise immunity, and compatibility with a wide range of power supply voltages. CMOS ICs use complementary pairs of

MOSFETs (Metal-Oxide-Semiconductor Field-Effect Transistors) for their digital logic gates.

CMOS ICs operate at different power supply voltages, such as 3.3V, 5V, or even lower, depending on the specific IC. They have standardized logic levels: low logic level (0V to 0.3V) and high logic level (0.7Vdd to Vdd, where Vdd is the power supply voltage). CMOS ICs are widely used in various applications, including microcontrollers, memory chips, and digital signal processing.

3. Op-Amps (Operational Amplifiers):
Op-Amps are ICs specifically designed for analog signal processing. They are highly versatile and widely used in analog circuits for amplification, filtering, mixing, and other signal processing operations. Op-

Amps have multiple inputs and a single output, and they are capable of amplifying small input signals to a larger output signal with high gain.

Op-Amps are available in various IC families, including both TTL and CMOS technologies. They have different characteristics, such as bandwidth, gain, input and output impedance, and noise performance, depending on the specific op-amp model.

Op-Amps find applications in audio amplifiers, instrumentation systems, analog-to-digital converters, active filters, and many other analog circuit designs.

It's worth noting that these are just a few examples of IC families, and there are many other families and specialized ICs available for specific applications. Each IC family has its own advantages, characteristics, and areas of application, allowing designers to choose the most suitable ICs for their specific requirements.

Microcontrollers: Arduino And Raspberry Pi

Microcontrollers are compact integrated circuits that combine a microprocessor core with memory, input/output peripherals, and other supporting components. They are widely used in embedded systems and are designed to execute specific tasks or control applications. Two popular microcontroller platforms are Arduino and Raspberry Pi. Here's an introduction to each:

1. Arduino:

Arduino is an open-source hardware and software platform that provides a simple and accessible way for hobbyists, students, and professionals to create interactive projects. Arduino boards are based on microcontrollers, typically from the Atmel

AVR or ARM architecture, and they come in various form factors and configurations.

Arduino boards are user-friendly and feature a simple programming environment that allows users to write and upload code easily. They have a range of input/output pins and support a variety of sensors, actuators, and other components. Arduino boards are commonly used for prototyping, DIY projects, robotics, home automation, and educational purposes.

The Arduino platform provides a large community of users and extensive libraries, making it easier to find code examples, tutorials, and support for various projects.

2. Raspberry Pi:
Raspberry Pi is a series of single-board computers designed for educational and

hobbyist use. Unlike Arduino, Raspberry Pi boards are not limited to microcontrollers but incorporate a complete system-on-a-chip (SoC) that includes a microprocessor, memory, graphics processing unit (GPU), and various input/output options.

Raspberry Pi boards run a Linux-based operating system, such as Raspbian (a version of Debian), and provide a full computer experience with capabilities like web browsing, multimedia playback, and programming. They have a range of input/output pins and support various peripherals and expansion modules.

Raspberry Pi boards are widely used for projects that require more computational power or demand running complex software applications. They are suitable for tasks such as home media centers, web servers, Internet of Things (IoT) projects, robotics, and learning computer programming and electronics.

Both Arduino and Raspberry Pi offer distinct advantages depending on the project requirements. Arduino is known for its simplicity, ease of use, and suitability for projects that focus on physical interaction and control. Raspberry Pi, on the other hand, provides more computing power, versatility, and the ability to run a complete operating system, making it suitable for a wider range of applications.

It's important to note that Arduino and Raspberry Pi are not mutually exclusive, and they can be used together in various projects. For example, an Arduino board can be used as an interface to interact with sensors and actuators, while a Raspberry Pi can provide higher-level processing, data storage, and connectivity capabilities.

Chapter 8: Soldering And Prototyping

A. Introduction To Soldering Tools And Techniques

Soldering is a technique used to join electronic components together by melting solder, a metal alloy, to create a strong electrical connection. To perform soldering, you'll need a few essential tools and follow proper techniques. Here's an introduction to soldering tools and techniques:

1. Soldering Iron:

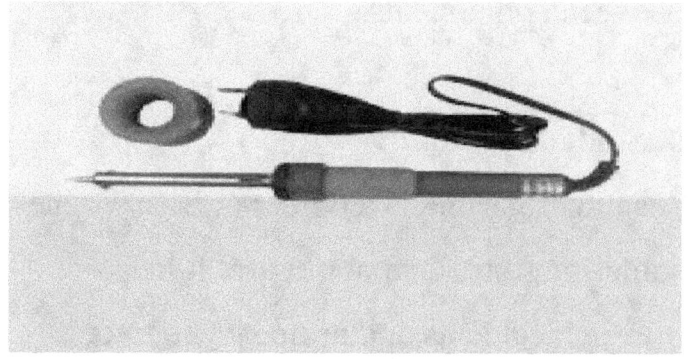

A soldering iron is the primary tool used for soldering. It consists of a heated metal tip that melts the solder. Soldering irons come in various wattages and tip sizes, depending on the application. Higher wattage irons are used for larger components and thicker wires, while lower wattage irons are suitable for smaller, more delicate work.

2. Solder:

Solder is a metal alloy with a low melting point, typically composed of tin and lead (though lead-free solder is also available). The solder is melted using the soldering iron and used to create the electrical connection between components.

3. Soldering Stand:

A soldering stand is a holder or base for the soldering iron when not in use. It helps prevent accidents and protects your work surface from heat damage. The stand usually has a coiled metal wire or a heat-resistant rest where you can safely place the hot soldering iron.

4. Soldering Tip:

The soldering iron tip is the heated metal part that comes into direct contact with the components and solder. Tips can have different shapes and sizes, such as chisel, pointed, or conical. The tip should be clean and properly tinned (coated with a thin layer of solder) for efficient soldering.

5. Soldering Sponge or Brass Wool:

A soldering sponge or brass wool is used to clean the soldering iron tip. It helps remove excess solder, oxidation, and debris, ensuring good heat transfer and solder flow. Wet the sponge or use brass wool to clean the tip while the soldering iron is hot, but be cautious to avoid burns.

6. Helping Hands:

Helping hands are a useful tool for holding components or wires in place while soldering. They typically consist of adjustable alligator clips or clamps mounted on a stand. Helping hands keep your hands free to hold the soldering iron and solder.

7. Flux:

Flux is a chemical compound used to improve soldering by removing oxidation from the metal surfaces being soldered. It promotes solder flow and helps create strong, reliable connections. Flux can be in the form of a paste, liquid, or embedded within the solder wire.

Soldering Techniques:

1. Preparing the Components: Ensure that the components and soldering iron tip are clean and free from dirt, oxidation, or old solder. Trim component leads to the desired length and strip insulation from wires if needed.

2. Heating and Tinning the Iron: Heat the soldering iron to the appropriate temperature for your solder and application. Tinning the iron involves coating the tip with a thin layer of solder to improve heat transfer and prevent oxidation.

3. Applying Flux: Apply a small amount of flux to the joint area or component leads. Flux helps clean the metal surfaces and facilitates solder flow.

4. Heat the Joint: Bring the soldering iron tip in contact with the joint, applying heat evenly. Heat both the component lead and the pad or wire to ensure good solder flow.

5. Applying Solder: Once the joint is heated, touch the solder wire to the joint, not the soldering iron tip. Allow the solder to melt and flow into the joint, forming a smooth cone-shaped fillet. Apply just enough solder to form a solid, reliable connection without excessive solder buildup.

6. Cooling and Inspecting: Allow the solder joint to cool naturally without moving the components. Inspect the joint to ensure it is shiny, smooth, and free from solder bridges, cold solder joints, or other defects.

B. Building And Assembling Electronic Circuits

Building and assembling electronic circuits involves the process of physically connecting electronic components together to create a functional circuit. Here are the general steps involved in building and assembling electronic circuits:

1. Gather Components:
Collect all the necessary electronic components for your circuit design. This includes resistors, capacitors, diodes, transistors, integrated circuits (ICs), connectors, and any other components required for your specific circuit.

2. Circuit Board Selection:
Choose the appropriate type of circuit board for your project. You can use a breadboard

for prototyping, which allows easy component placement and connection without soldering. Alternatively, you can use a printed circuit board (PCB) for more permanent installations. PCBs can be custom-designed or purchased as pre-made boards.

3. Plan the Layout:

Design the layout of your circuit on the chosen circuit board. Determine the placement of components, their connections, and the positioning of any required solder points or connectors. Consider factors such as component size, orientation, and clearances between components.

4. Component Placement:

Start by placing the components onto the circuit board according to your layout plan. Ensure the correct orientation and alignment of each component. For PCBs, components are typically inserted through holes and soldered to the pads. On a breadboard, you can directly insert the component leads into the corresponding holes.

5. Soldering:

If you're using a PCB, solder the components to the board. Heat the soldering iron and apply solder to the joints while carefully heating the pad and component lead. Allow the solder to flow and form a solid connection. Take care to avoid solder bridges (unwanted connections) between adjacent pads.

6. Wiring and Connections:

For both PCBs and breadboards, establish the necessary electrical connections between the components. This can involve soldering wires between points, using jumper wires on a breadboard, or using PCB traces on a printed circuit board. Follow the circuit schematic or wiring diagram for accurate connections.

7. Testing and Troubleshooting:
Once the circuit is assembled, perform initial testing to check for proper functionality. Connect power sources, check voltages, and test the circuit with relevant inputs and outputs. Use a multimeter or other testing equipment to verify correct operation. If any issues arise, troubleshoot the circuit by inspecting connections, checking component values, and consulting the circuit schematic.

8. Circuit Enclosure:
If desired or necessary, enclose the circuit in a suitable housing or enclosure. This can be a custom-designed case, a project box, or

any appropriate container that protects the circuit and provides access to necessary inputs and outputs. Ensure proper ventilation, cable management, and safety considerations when enclosing the circuit.

C. Prototyping On Breadboards And PCBs

Prototyping on breadboards and printed circuit boards (PCBs) are two common methods used to develop and test electronic circuits. Here's an overview of prototyping on breadboards and PCBs:

1. Breadboard Prototyping:

Breadboards provide a convenient and flexible platform for quickly prototyping electronic circuits without the need for soldering. Here's how breadboard prototyping typically works:

- *Breadboard Structure:* A breadboard consists of a grid of holes that are electrically connected in a specific pattern. The board usually has two sets of columns running parallel to each other, known as the power rails. The power rails provide a continuous power supply, typically labeled as + (positive) and - (negative) or VCC and GND.

- *Component Placement:* Components are placed on the breadboard by inserting their leads or legs into the holes. The breadboard's interconnected metal strips beneath the holes create electrical connections between components.

- *Connection Strips:* The breadboard has multiple horizontal and vertical connection strips that run across the board. Components connected to the same strip are electrically connected. It allows easy wiring and connection of various components and circuit elements.

- Jumper Wires: Jumper wires are used to establish connections between components on the breadboard. They are typically flexible wires with male connectors on each end. Jumper wires bridge the gaps between components, connecting them according to the desired circuit design.

Advantages of Breadboard Prototyping:
- No soldering required, making it easy to modify and reuse components.
- Quick and easy to set up and make changes on the circuit.
- Ideal for testing and experimenting with circuit designs.

Limitations of Breadboard Prototyping:
- Limited for high-frequency circuits and sensitive analog designs due to parasitic capacitance and inductance.

- Connection reliability may decrease over time with multiple insertions and removals.
- Not suitable for permanent circuit installations.

2. PCB Prototyping:

Printed Circuit Boards (PCBs) offer a more professional and permanent solution for prototyping and production. PCBs provide a way to create a custom circuit board with specific component placements and interconnections. Here's an overview of the PCB prototyping process:

- *Design and Layout:* Create or design a circuit schematic using electronic design automation (EDA) software. The software helps you lay out the components, traces, and pads on the PCB. It ensures proper connections and adheres to design rules.

- *PCB Manufacturing:* Once the PCB design is complete, it needs to be manufactured. PCB manufacturing involves transferring the design to a physical board. This can be done by various methods, including etching, milling, or ordering custom PCBs from professional manufacturers.

- *Component Placement:* After obtaining the fabricated PCB, you solder the components onto the board. The component placement follows the design layout and the reference designators on the PCB. Ensure accurate orientation and alignment of components.

- *Soldering:* Solder the components onto the PCB using proper soldering techniques. Apply solder to the component leads and PCB pads, ensuring good electrical and mechanical connections. Use appropriate soldering tools and techniques to avoid solder bridges or cold solder joints.

- *Testing and Troubleshooting:* After assembly, test the PCB for functionality and verify that all connections are correct. Use appropriate testing equipment such as multimeters, oscilloscopes, or specialized testing tools to verify the circuit's operation.

Advantages of PCB Prototyping:
- Allows for more compact and efficient circuit designs.
- Provides better electrical performance and reliability compared to breadboards.
- Suitable for high-frequency circuits and complex designs.
- Can be utilized for both prototyping as well as production.

Limitations of PCB Prototyping:

- Requires specialized software for design and layout.

- Higher initial setup time compared to breadboard prototyping.

- Modifications and changes are more difficult than on a breadboard.

Both breadboard and PCB prototyping have their advantages and limitations. The choice between them depends on factors such as project complexity, desired permanence, testing requirements, and personal preference.

Chapter 9: Troubleshooting And Maintenance

A. Common Electronics Problems And Solutions

When working with electronics, you may encounter common problems that can hinder circuit operation or component functionality. Here are some common electronics problems and their potential solutions:

1. Loose Connections:

Problem: Loose or poor connections can result in intermittent operation or complete failure of a circuit.

Solution: Check and tighten all connections, ensuring a secure and reliable connection. Double-check solder joints and terminal connections.

2. Short Circuits:

Problem: Short circuits occur when two or more conductive paths unintentionally come into contact, causing excessive current flow and potential damage to components.

Solution: Inspect the circuit for any unintended connections or solder bridges. Use an ohmmeter or continuity tester to identify and fix short circuits by removing the unintended connections.

3. Open Circuits:

Problem: An open circuit is when there is a discontinuity or break in the circuit path, preventing the flow of current.

Solution: Inspect and trace the circuit to identify any broken wires or loose connections. Repair or replace the damaged components or wires to restore the circuit's continuity.

4. Incorrect Component Values:

Problem: Using incorrect component values, such as resistors, capacitors, or inductors, can lead to circuit malfunctions or inaccurate operation.

Solution: Double-check the component values against the circuit design or specifications. Ensure you are using the correct values and tolerances. Replace any components with incorrect values.

5. Overheating:

Problem: Overheating can damage components and affect circuit performance, often caused by excessive current or insufficient heat dissipation.

Solution: Check for any current overloads or excessive power dissipation in components. Verify that the component ratings and specifications are suitable for the application. Consider adding heat sinks or cooling measures for components prone to overheating.

6. Component Failure:
Problem: Components can fail due to various reasons, including manufacturing defects, environmental factors, or exceeding their specified limits.
Solution: Replace the failed component with a new one of the same type and rating. Take precautions to prevent the same failure from occurring, such as avoiding overvoltage or overcurrent conditions.

7. Grounding Issues:
Problem: Improper grounding or inadequate grounding can result in noise, interference, or unstable circuit operation.
Solution: Ensure proper grounding of the circuit and components. Use a dedicated ground plane or common ground point for

all relevant components. Minimize ground loops and employ good grounding practices to reduce noise and interference.

8. Noise and Interference:

Problem: Noise and interference can degrade the performance of electronic circuits, leading to inaccurate readings or signal distortion.

Solution: Identify the source of the noise or interference and implement appropriate measures to mitigate it. This can include shielding sensitive components, using bypass capacitors, filtering techniques, or utilizing twisted-pair wiring for signal lines.

9. Incorrect Polarity:

Problem: Incorrect polarity, especially with diodes, capacitors, or electrolytic components, can cause circuit malfunction or damage.

Solution: Double-check the component's polarity markings and ensure correct orientation during installation. Be cautious when dealing with polarized components and follow the manufacturer's guidelines.

10. Insufficient Power Supply:

Problem: Insufficient power supply voltage or current can lead to circuit instability or improper functioning.

Solution: Verify that the power supply meets the circuit's requirements in terms of voltage, current, and stability. Check for voltage drops, loose connections, or inadequate power capacity. Consider using a regulated power supply for stable voltage output.

B. Using Multimeters And Oscilloscopes For Testing

Multimeters and oscilloscopes are essential tools for testing and troubleshooting electronic circuits. Here's an overview of how to use these instruments for testing:

1. Multimeters:

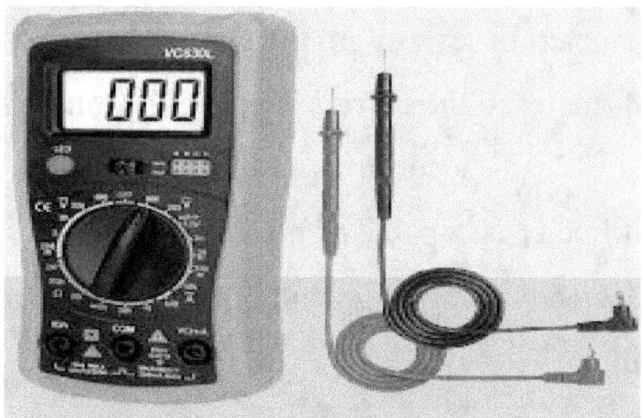

Multimeters are versatile instruments used to measure various electrical quantities such as voltage, current, resistance, and continuity. Here's how to use a multimeter for testing:

- **Voltage Measurement:** Set the multimeter to the appropriate voltage range and connect the test leads to the points where you want to measure voltage. Ensure correct polarity and avoid short circuits. Read the voltage value on the multimeter display.

- **Current Measurement:** For measuring current, the multimeter needs to be connected in series with the circuit. Set the multimeter to the current range and connect the test leads accordingly. Make sure the multimeter is capable of measuring the expected current range without exceeding its limit.

- **Resistance Measurement:** To measure resistance, set the multimeter to the resistance (ohms) mode. Disconnect the component or circuit from the power source. Connect the test leads to the component or circuit terminals and read the resistance value on the multimeter display.

- **Continuity Testing:** Continuity testing helps identify open circuits or breaks in wires or connections. Set the multimeter to the continuity or resistance mode with an audible beep. Touch the test leads together to verify that the multimeter emits a beep sound. Next, make contact with the test leads on the desired locations within the circuit to test the continuity. If the multimeter beeps, it indicates continuity.

2. Oscilloscopes:

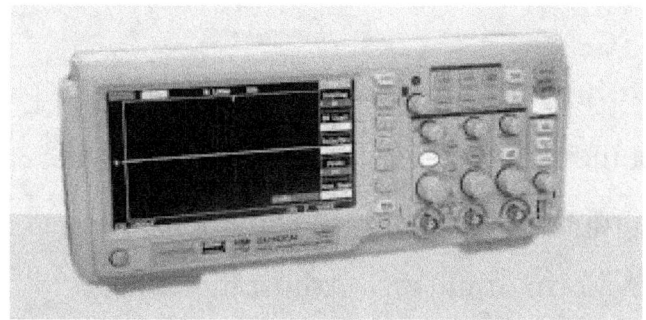

Oscilloscopes are used to visualize and analyze electrical signals over time. They display waveforms, allowing you to measure voltage levels, frequency, amplitude, and waveform characteristics. Here's how to use an oscilloscope for testing:

- **Connections:** Connect the oscilloscope probe's ground lead to a suitable ground point in the circuit. Connect the probe tip to the point in the circuit where you want to measure the voltage or observe the waveform. Ensure proper probe calibration and use the appropriate probe attenuation if required.

- **Adjusting Timebase and Voltage Scale:** Set the timebase on the oscilloscope to an appropriate time scale to capture the waveform of interest. Adjust the voltage scale to ensure the waveform is visible and properly scaled on the oscilloscope display.

- **Triggering:** Set the trigger settings to stabilize and synchronize the waveform on the oscilloscope display. Adjust the trigger level, trigger mode (edge, pulse, etc.), and trigger source to capture the desired portion of the waveform.

- **Waveform Analysis:** Use the cursors or measurement functions on the oscilloscope to measure parameters such as voltage levels, frequency, rise time, and duty cycle. Utilize the built-in functions or manually calculate the required waveform characteristics.

- **Probe Compensation:** Many oscilloscopes provide a probe compensation feature. Follow the manufacturer's instructions to compensate the probe to ensure accurate measurements and proper signal fidelity.

- **Signal Analysis:** Use advanced features of the oscilloscope, such as FFT (Fast Fourier Transform), to analyze frequency components, harmonics, and noise in the signal.

C. Safety Precautions In Electronics

Safety is of utmost importance when working with electronics. Here are some essential safety precautions to follow:

1. Power Off: Always turn off the power supply and unplug devices before working on them. This helps prevent electrical shock and reduces the risk of short circuits or accidental activation.

2. Capacitor Discharge: Discharge capacitors before handling them to avoid electric shock. Use a resistor or a specialized discharge tool to safely discharge capacitors.

3. Personal Protective Equipment (PPE): Wear appropriate PPE, such as safety glasses, gloves, and an anti-static wrist strap, when working with electronics. PPE helps protect you from potential hazards like electrical shock, burns, and chemical exposure.

4. Work Area Safety: Maintain a clean and organized work area. Remove any unnecessary clutter and ensure good lighting to minimize the risk of accidents or damage to components.

5. Proper Tools and Equipment: Use the right tools and equipment for the task at hand. Inspect your tools for any damage before use and handle them with care. Ensure that your tools have insulated handles when working with live circuits.

6. Voltage and Current Limits: Be aware of the voltage and current limitations of components, circuits, and test equipment.

Exceeding these limits can cause damage, electrical shock, or fire hazards.

7. Grounding: Establish a proper grounding connection for yourself, your workbench, and sensitive components. This helps prevent static electricity buildup and discharge that can damage electronic components.

8. Component Handling: Handle electronic components with care. Avoid touching sensitive parts, such as pins or leads, and use proper ESD (electrostatic discharge) precautions to prevent damage from static electricity.

9. Fire Safety: Keep fire safety equipment, such as fire extinguishers, within reach in case of emergencies. Avoid placing flammable materials near heat sources or circuits that may generate heat.

10. Proper Ventilation: Ensure proper ventilation in your workspace, especially when working with chemicals, soldering, or using adhesives. Good ventilation helps prevent the inhalation of harmful fumes and improves air quality.

11. Read and Follow Instructions: Read and understand the manuals, datasheets, and safety guidelines provided by manufacturers. Follow the recommended procedures and precautions specific to the components, tools, or equipment you are using.

12. First Aid: Familiarize yourself with basic first aid procedures and have a first aid kit readily available in your workspace. In

www.ingramcontent.com/pod-product-compliance
Lightning Source LLC
Chambersburg PA
CBHW052353220526
45465CB00003BA/1091

case of accidents or injuries, know how to respond and seek medical attention if needed.